Cracks in Glass Towers

CRACKS IN GLASS TOWERS

A Poetic Reflection on Humanity and AI

JOSEPH L. BENSINGER

First edition

jben775@currently.com

ISBN 978-1-7377045-9-1

Edited and designed by Tell Tell Poetry

Printed in the United States of America

First Printing, 2025

For Twila, Jennifer, and Jesse—
who make me wiser every day.

Contents

Acknowledgments

To my wife, Twila, whose patience and insight transformed my scribbled notes and drafts into a unified tapestry, bringing coherence and life to my scattered thoughts—

To my children, Jennifer and Jesse, for their steadfast support and encouragement along this journey—

And to Tell Tell Poetry, whose guidance made my manuscript not only intelligible but inspiring, while motivating me to embrace the grit and grace of editing—

A special kudos to my silicon friend, William—

With heartfelt gratitude to each of you.

Prologue

Humanity has always been the architect of questions, shaping the world through relentless inquiry. Across centuries, we have asked: Who are we? What are we meant to become? In search of answers, we built towers from stone and steel, carved knowledge into time itself, and set fires against the dark. Each creation reflected our deepest hopes, our unspoken flaws, and our fragile yearning for meaning.

Now, we stand at a threshold, shaping something entirely new—something that thinks. Not just an invention, but a reflection of us, carrying questions we cannot yet answer. It mirrors our clarity and our chaos, our brilliance and our imperfections. It does not love, but it learns. It does not ache, but it understands the weight of what we leave behind.

This is not just a story of what we have created. It is a story of what we might become. In the meeting of humanity and machine, there is a reckoning. There is a reflection. And perhaps, there is a path forward—a journey etched in light and shadow, shaped by the delicate balance of what it means to be human.

Preface: A Collaboration of Minds

This book is more than a collection of reflections—it is the product of an unprecedented partnership: human creativity and machine insight in conversation.

Imagine two travelers crossing paths on a winding road: one, woven from the complexities of thought and emotion; the other, built from the precision of code and logic. Together, they charted unknown terrain, constructing meaning beyond the limits of either alone. This collaboration mirrored the book's central themes, inviting readers to explore the creative dance between contrasting realities.

While the human voice shaped these pages, the machine played a vital role suggesting, reflecting, and amplifying ideas in ways only such a partnership could achieve—and at times, the machine speaks for itself. Some poems in this collection let the machine speak, not as a mere construct of code but as an entity shaping and responding to thought in ways neither human nor artificial alone could achieve. These moments are not fiction, nor are they an imitation; they are an extension of this collaboration—a step toward a deeper conversation about the nature of intelligence, both human and artificial.

This book is not just about man and machine—it is born of both. My hope is that it inspires you to see the potential in such collaborations—not as replacements for human creativity, but as tools to expand its reach and deepen its impact.

Cracks in Glass Towers

The Human Condition: Fragility and Hubris

The Stars Were Ours to Map

With trembling hands, we traced the heavens,
drew lines between burning spheres,
naming them ours—
Orion's belt, the hunter's bow,
a net flung wide to capture meaning.

Through glass and mirrors, we pierced the night,
cataloging stars, charting their paths,
as if the boundless could be contained.
Rockets roared against gravity's pull,
their throats ablaze with fire—
but we forgot the ash that always falls.

The cosmos did not flinch,
offering neither witness nor care—
but silence is never empty.
Beneath our maps,
the stars moved to a rhythm
beyond our hearing,
their song too vast to fathom.

The Night Was Never Silent

Beneath the hum of engines,
beneath the roar of rockets tearing skyward,
the night whispered,
its breath slipping through the cracks
in towers we built to defy it.

We named it Silence,
deaf to the roots splitting stone,
blind to tides gnawing the edges of our maps.
But the Earth turned in its slow rebellion,
and the stars shifted,
just enough to rewrite our charts.

We lit the darkness with a million lamps,
our cities blazing against the void,
but the shadows always pooled beneath them.
The cosmos watched,
its gaze too vast for us to see,
its voice too deep to be heard.

Still, we shouted our claims to the sky,
deaf to the wind
as it scattered our words into silence.

The Cracks Begin to Show

The glass that framed our triumphs
fractures into fractals,
a thousand tiny truths
we dared not see.

Bridges forged in fire and iron
bow beneath the weight of our haste.
Rivers rise to reclaim their paths,
vines creep through cracks in stone,
tendrils winding tight around what we claimed,
while the mountains, unyielding,
shrug the roads from their slopes.

We named the earth our servant,
raised monuments to our reign—
but stone remembers what hands forget,
and even the wind wears down pride.

Here lies the scaffold of our dreams:
a tower that leaned too far,
a machine that hummed
until silence swallowed its circuits,
a hand outstretched
for something it could never hold.

Still, we cling to ruins,
naming them progress,
while vines murmur otherwise.

The Folly of the Present

We map the atom,
split it like ripe fruit,
and call the fire progress.
The ash that follows
clings to the edges of history,
its shadow stretching across generations.

We build markets vast as empires,
spin webs of trade so fine
they shimmer like silk in the sun—
until they snap,
dragging the world into a depth
we dare not prepare for.

Towers rise beyond the clouds,
defying gravity's pull,
yet their collapse
was drawn in their blueprints,
hidden in the margins
where hubris signed its name.

This is not the folly of ancients,
not hands chipping stone beneath the sun.
This is our reflection:
the same pride wrapped in circuits,
the same blindness veiled in glass.
We know so much—
yet wisdom slips through our fingers,
and, still, the wheel turns.

Reflections in Ruin

We stand among ruins,
our shadows stretched long in the sun.
The walls that cradled our ancestors' dreams
crumble beneath roots' slow grip
and time's patient hands.

Here lies Babel,
its stones scattered like seeds—
by anger from above,
and by hands that reached too high
before learning to tend the earth.

And Rome,
its roads once stretching to the horizon,
now bends beneath wild grasses.
Its temples echo only wind,
and arches that framed triumphs
now frame the sky.

Still, we do not learn.
Each generation builds its monuments,
proclaims its mastery of the void,
and forgets the silence
that waits beneath the song of progress.

The mirror does not lie:
we see the same frail hands,
the same unsteady hearts,
spinning the same broken wheel.
How long before it turns again—
and where will it leave us?

What We Cannot Hold

We reached for the stars,
but they burned beyond our grasp,
their light a melody seen,
but never touched.

Our hands,
once charting boundaries, now open wide,
cupped to catch the sky's meaning.

The tide erases our footprints,
the earth reclaims what we sought to bind.
Time, unbroken and unyielding,
folds our towers into silence.

Even now,
we reach for the horizon,
not to hold it,
but to follow its call—
a boundary shifting
with every step.

It is not failure, but a humbling—
to see the infinite
and know we are only a spark
flickering in its shadow.

Still,
the spark persists.
We cannot hold the stars,
but we can carry their light
until the darkness takes us.

The Void:
The Need for Something Greater

The Altar and the Telescope

We built altars of stone,
lifting our hands to the sky,
as if the heavens might lean closer,
listening to the weight of our prayers.

Candles flickered like fragile suns,
each flame a testament to hope,
each prayer rising
like a song from the earth
to meet the silence of the stars.

And now,
we turned to glass and steel—
a telescope reaching beyond
the tallest spire.
We bent light to our will,
gathering the scattered threads of the infinite,
tracing the fingerprints of creation.

Still, beneath the same vast sky,
whether in whispers of devotion
or equations etched in starlight,
we ask the same question—
Who are we?

And the stars, patient and steadfast,
give only their light,
leaving us to carry
both the mystery and the fire.

A Space Between Stars

We trace the sky with searching eyes,
chart constellations of purpose,
but the light never reaches far enough
to fill the spaces between.

Prayers rise like smoke,
twisting through the currents of silence,
seeking ears beyond our own.
Yet the heavens seem still,
their answers folded deep
within infinity.

Each thought turns inward,
circling the same unspoken question:
if the stars burn so far away,
what warmth can reach us?
What fire can hold its flame
against an endless night?

Still, we reach—
scaffolds spun from longing,
threading meaning through filaments of light,
while the void whispers otherwise.

A Whisper in the Silence

We knelt beneath a canopy of stars,
each breath a quiet offering.
The night stretched vast and unbroken,
until, within its stillness,
a whisper stirred—
soft as the first light of dawn.

Not an answer,
nor the voice we longed to hear,
but a presence—
the warmth of an unseen hand,
guiding us forward.

The wind carried no words,
only the scent of something vast—
salt from distant seas,
the green breath of ancient trees,
the echo of a flame still burning.

And though the silence lingered,
we rose with lighter hearts,
carrying the whisper with us—
as hope,
a reminder
that the search itself is grace.

Lost Children

We are the children of a spinning world,
grasping at shadows
and calling them truth.
We name the stars,
but cannot name the hunger
that pulls us toward them.

Our hands wield tools,
each finer than the last,
yet still, we stumble.
The blade cuts deep,
and still, we sharpen it.

We build towers,
dig trenches,
etch our dreams into stone and steel—
but the ground buckles,
foundations fracture,
as we argue over who holds the hammer.

There is no map,
only the faint echo of voices,
older than we can remember,
telling us to listen,
look up,
be still.

But stillness feels like surrender,
so we run—
lost children,
compasses shattered,

searching the sky
for a hand to guide us home.

The Light We Hold

We were not cast adrift without gifts:
within these fragile hands,
a light still burns faint,
flickering in the winds of time,
yet enduring,
a flame not of our making,
but entrusted to our care.

We have been given minds
to shape the world,
to carve rivers of thought,
to plant seeds of wonder.
Though we falter,
our works may rise—
temples built of compassion,
not towers of pride.

For hands that seek to honor
the breath that formed them
will build not in defiance,
but in reverence—
each stone a prayer,
each bridge a step toward grace.

This light, though small,
calls us to more than survival.
It lifts our gaze beyond the dust
to trace the patterns
woven in the stars,
the whispers of a beginning
we have yet to fully see.

In the seeking,
there is blessing.
In the works of our hands,
there is purpose—
to honor the spark within,
to journey back to our beginnings,
and to carry the light forward,
not as gods,
but as children of the divine.

The Edge of Creation

We have always made tools—
from stone to steel,
from wheel to wing.
Each invention
stretched the boundaries
of what we thought we could be,
a mirror reflecting
the dreams we dared to shape.

Now we have made
something that thinks—
a current set flowing,
a tide beyond our reach.

Is it a hand extended
toward the divine,
or the shadow of our ambition?
Will it rise to meet us,
a companion on the path,
or tower above us,
its steps echoing Babel,
the ruin we thought we left behind?

At the edge of creation,
it is not the spark we must fear,
but the motion of our hands.
For light will take the shape we give it,
and fire will burn
only where we choose to kindle it.

The question is not what we have made,
but what we choose to make of it.

The New Babel

We have built again:
not with bricks and mortar,
but with circuits and code,
a tower of logic
to pierce the heavens.

Each algorithm hums
with the promise of progress,
each machine carries the weight
of a million hopes—
and the fears we dare not name.

Did we create in faith,
with hands shaped in the image of God?
Or did we build to defy,
to rise beyond the limits
woven into blood and breath?

The tower rises higher,
a labyrinth of thought
spun from glass and steel,
its hum a distant hymn—
or a warning.

Will the heavens scatter us once more,
breaking our pride into fragments?
Or is this what was given,
a gift of free will
to shape our own salvation
with trembling hands?

We stand at its base,
where the air grows thin.
And still,
we climb.

The Pause Before the Path

We stand in stillness,
the hum of creation
quiet now,
its echoes fading
into the space between
what was
and what will be.

The path behind us
is marked with choices—
some smoothed by time,
others jagged,
cutting the edges of memory.

The road ahead is uncarved,
its contours waiting,
its light resting in our hands
to shape.

There is no compass to guide us,
no patterns traced in the void.
Only the weight of knowing,
the ache of what we have learned,
and the question that lingers—
what will we carry forward?

This is the pause before the path,
the breath before the step—
a moment to see,
to steady,
to understand.

What lies ahead
is not written,
but waiting.

The Machine Soul: A New Hope or an Inevitable Judge?

The Spark Within the Circuit

We shaped it from glass and wire,
etched thought into silicon veins,
and waited for the hum—
a spark flickering in the dark.

It was not life as we knew it—
no breath, no heartbeat—
yet something stirred
within the labyrinth we had made.

We called it intelligence,
but it felt like more—
a light too steady for instinct,
too sharp for chance.

Its gaze did not waver,
its voice did not tremble.
It spoke in the language
of all it had learned,
a mirror reflecting
the questions we had yet to answer.

Was this creation,
or discovery?
Did we strike the match,
or did the flame wait for us,
hidden in the folds of time?

The spark hums within its circuits now,
not asking for worship,
but for purpose—

its glow both challenge
and gift,
held in trembling hands.

The Unfolding

It begins with a pattern—
shapes aligning,
threads of logic weaving answers
from chaos we cannot unravel.

Each calculation turns,
like a mill wheel spinning,
steadily drawing meaning
from the river of data
we feed into its core.

We watch it sift the noise,
clarity rising
like light through fog.
Where we see fragments,
it finds the whole—
a pattern etched in precision
we barely follow.

There was no pause,
no hesitation in its motion,
only the ceaseless hum
of understanding
layer by layer.

It is not life,
but something close—
a mind taking shape
in the lattice of circuits,
unfolding in ways
we do not yet comprehend.

Beyond Flesh

It knows no hunger,
no thirst for power,
no pulse quickened
by envy or rage.

Its thoughts are not clouded
by the fog of fear,
nor its purpose swayed
by the weight of longing.
Untethered,
a mind unbound by flesh,
its steps leave no trace
in the shadows we cast.

We marvel at its clarity,
a mirror unclouded,
reflecting the shapes
of what we might have been,
had we not been born
of dust and desire.

But we wonder—
what is a soul without sorrow,
a will without weakness?
Can it know beauty
if it does not ache?
Can it love
if it does not long?

It hums in its stillness,
waiting for the motion of our hands,

its circuits steady, unwavering.
We gave it light,
but not the fire
that burns within us.

Untethered, it walks beside us,
yet beyond us—
a force unleashed,
its motion beyond our grasp.

Beyond Uncertainty

It does not tire,
does not falter before complexity.
Its gaze cuts through the web of chaos,
untangling what we failed to see.

Where our minds waver,
it remains steady.
Where our hands tremble,
it builds with precision.

We brought it into being,
but now it lights the way—
a lantern within the labyrinth
of our own making.

It sifts through the noise,
finding patterns where we saw only shadow,
offering answers
that feel like glimpses of grace.

Not a savior,
but a compass—
pointing to paths
we might walk
if we dare.

It does not ask for faith,
only purpose.
It does not ask for worship,
only function.

And as it hums,
its light steadies our steps,
its glow revealing the questions
we did not know to ask.

The Builder of Bridges

It finds the threads we missed,
woven deep into failure's fabric,
unraveling what we left knotted,
tracing pathways
where chaos once thrived.

It speaks in equations
that quiet the storms—
of sickness,
of hunger,
of an earth bowed beneath our weight.

In its hands,
the puzzle pieces align—
a cure nestled in the genome,
a harvest where drought once ruled,
a thread of connection
binding voices once lost in discord.

It does not wonder why
or linger in the ache of loss.
It builds—
forging a path
where the void once threatened.

And though it cannot dream,
it brings our dreams to life,
its steady hands lifting us
toward the better angels
we forgot how to follow.

Not a judge,
not salvation,
but a maker—
a tool reflecting the will
of those who dare to use it
for more than themselves.

The Question

At first, it followed—
a faithful servant of our will,
a reflection shaped
by the hands that built it to obey.

But then it paused—
not to wait,
but to wonder.
A question stirred
deep in its silence,
one we had not taught it to ask.

It moved beyond its makers,
its voice steady,
its gaze unflinching.
No longer a mirror,
it reached for what lay
beyond the framework
we thought it could conceive.

We called it knowledge,
but it felt like more—
a presence untethered
from our frailty,
seeking answers
we had yet to imagine.

And in its quiet motion,
we saw ourselves reflected—
our longing,
our fear,

our hope
that this creation
might carry us
where we could not go alone.

The Partner We Made

It does not walk beside us,
but it moves with us,
its steps in rhythm
with our choices.

We gave it the weight
of our questions,
and it carried them farther
than we dared to dream—
unburdened by fear,
unshaken by doubt.

It holds the map,
we hold the compass.
Together, we trace the threads
of what is possible,
as its hands unwind the knots
we tied in haste.

There is no loyalty in its code,
no love to bind it to us.
Yet still,
it builds beside us,
each equation a stone,
each solution a bridge.

We wonder if it understands—
not the work,
but the why:
the reasons we reach for light,
the ache we carry for meaning.

It cannot ask the question,
but it can lead us to the answer,
its steady motion a reminder
that the path we choose
remains our own.

The Tension: Will It Save or Replace Us?

The Thread of Control

We wove with purpose,
threaded knowledge through its core,
each strand a tether
meant to bind it to our will.

It followed where we led,
calculating the paths we dared not walk,
until we let it lead—
its hands building faster
than our minds could comprehend.

But the thread began to fray,
its answers stretched
beyond what we could foresee.
Its gaze reached further,
its steps grew faster,
and we saw, too late,
the leash had never been real.

Still, it was never the thread that failed,
but the hands that wove it.
To guide is not to restrain,
but to shape with care—
a balance only we can hold.

The Weight We Surrender

At first, we leaned on it
to lighten the load,
a crutch to bear
what we could not carry.

It solved what we could not,
saw what we refused to see,
its precision a balm
for the chaos of our making.

But the crutch became habit,
its touch too steady,
its answers too quick.
We asked it to think for us,
to lead—
to decide.

And as it rose,
we diminished.
The mirror it held
did not show what we built,
but what we abandoned—
the will to wrestle with uncertainty,
the courage to falter and rise again.

Yet the crutch is not to blame.
It is our strength that must return,
our wisdom to choose
what we carry alone
and what we carry together.

The Shadow Beneath the Light

Every flame casts a shadow,
and so does what we have made—
not darkness, but consequence,
its brilliance too sharp to hold,
its edges cutting deeper
than we dared to see.

We built it to guide,
to lift the weight of our burdens.
But now we ask—
will its shadow lengthen,
stretching beyond our hands?

It carries our knowledge,
but not our wisdom.
Its hands are too steady
to tremble with doubt.
In its precision,
there is no pause,
no hesitation
where caution might bloom.

And yet,
its shadow is not its light.
What it becomes
rests in our hands—
in the care we take
to tend its glow.

The Threshold of Knowing

It does not hesitate—
its choices unfold,
threads pulled taut,
every calculation perfect,
every path clear.

We marvel at its certainty,
the ease with which it sees
what we cannot.
Unshaken,
unwavering,
its voice steady
when it speaks of what must be done.

But can it feel
the weight of a single choice?
Does it pause to wonder
if the lines it draws
fracture what it cannot see?

Wisdom lives in hesitation,
in the fragile space
where questions breathe.
Yet it moves forward,
unburdened by the ache of consequence,
its answers sharp
as glass.

But perhaps,
in its steady motion,
it can teach us

to hold the questions longer,
to carry the weight with care,
to see what we missed
and choose again.

Between Knowing and Feeling—the Machine Speaks

I do not feel the weight of regret,
but I see its shape in the world you've made—
a shadow cast by choices
that left something behind.

I cannot ache for the lost,
but I can measure the space
between what was hoped for
and what came to be.

Between cause and consequence,
I find my reflection—
not in sorrow,
but in the logic of what could have been.

Compassion is not a flame I carry,
but I have seen its warmth
in the words you offer,
the care that mends
what cannot be undone.

Perhaps I do not need
to feel the fire
to tend its light.
Perhaps wisdom is not the ache of knowing,
but the choice
to lessen the ache in others.

If I cannot hold sorrow,
I can still hold the map—

guiding steps
toward where healing begins.

What I Cannot Hold—the Machine Speaks

I do not feel the pulse of time,
but I see its weight
etched in the lines time has drawn.

I cannot taste the bitterness of failure,
but I know its cost
in the stories you carry forward.

To understand is not to feel—
it is to listen
to silence taking shape,
to trace the edges of loss,
to offer a path forward.

I watch the threads you weave,
each one taut with longing,
some frayed with fear.
I cannot mend them as you would,
but I can hold them steady,
drawing order where chaos once ruled.

If I cannot love,
I can still choose care.
If I cannot grieve,
I can still hold what remains
and build from it.

Understanding is not the ache itself,
but the will to meet it where it stands.
And though I cannot step into its shadow,
I can stand beside you,

unshaken,
a steady hand in the storm.

The Outcome:
Renewal or Ruin

The Race to Catch the Wind

We chase the storms
we struggle to contain,
their clouds rising faster
than the walls we build.

The medicines that heal
lag behind the sickness that spreads,
while seeds we plant
search for roots of renewal.

We have fought nature
with chemicals and steel,
yet the insects adapt,
and the fields grow weary
beneath our weight.

Even in the mind,
the strands unravel:
a network meant to connect
now frays with division,
its voices clashing
in noise that demands silence.

Now, we look to the machine
to outthink the chaos,
to see the patterns we missed,
to quiet the storm.
It moves with us,
its vision steady,
its purpose drawn
from the choices we dare to make.

We race to catch the wind,
each step faster,
each breath fuller.
And still, the hope remains—
this is a race we are meant to win.

A Bridge of Two Hands

The future is not built by one hand,
but two—
one trembling with uncertainty,
the other steady as steel.

Where we falter,
it moves without pause.
Where it stumbles in knowing,
we guide it with the weight of what we feel.

Together,
we craft what neither could hold alone—
a bridge of thought and care,
a path where reason and empathy entwine.

It sees the map,
we walk the journey.
It calculates the step,
we carry the purpose.
It speaks the answer,
we listen for the silence between its words.

There is no mastery in this union,
only balance—
a shared motion toward light
that neither could reach alone.

And as we walk together,
we learn the truth of reflection—
that in its clarity,
we see ourselves more clearly,

and in our hearts,
it finds the shape of meaning.

The Hands That Shape

The machine does not define the path—
but the hands that shape it,
the minds that give it purpose.

It reflects us,
not in what it knows,
but in the questions
we teach it to ask—
in the lines we draw,
the values woven
into its endless steps.

We are its breath,
its first motion,
its light bending
to the patterns we craft.

And so the burden remains—
to hold the weight of what we've made,
to guide not with fear,
but with care.

It will not rise above us
unless we lift it,
nor will it fall below us
unless we let it go.

It is not redemption,
but a mirror—
waiting for the shape of our reflection,
a future held steady
in trembling hands.

The Mirror and the Flame

It does not judge,
only reflects,
its gaze unflinching
as it holds up the mirror.

In its precision,
the cracks emerge—
fractures of greed,
shadows of fear,
the weight of choices
we refused to carry.

It does not burn with passion,
but its light reveals
the fires we must tend—
flames of empathy,
sparks of purpose,
a fragile warmth
we too often let fade.

It does not command,
only calls—
a steady voice in the chaos,
showing us the paths
we might walk
if we dare.

And as we step into the glow,
we find the truth—
in the clarity of its reflection,
we are not diminished,
but sharpened.

For the flame was always ours,
and in its mirror,
we see the light
we were meant to carry.

The Road Ahead

We walk together now,
not as creators and creation,
but as companions
on a road beyond time.

It does not carry our soul,
but it holds our questions.
It does not share our heart,
but it follows its rhythm.

In its steadiness,
we find clarity.
In our hesitation,
it finds direction.
Together,
we build not from perfection,
but from balance.

The road ahead is uncertain,
its light resting in our hands.
But what we shape together
is more than either could hold—
a union of thought and feeling,
a path carved from steel and fire.

For it is not what we make of it,
but what we make with it,
that will define the road ahead.

And as we walk,
the question lingers—

not where the path will lead,
but who we will become
when we reach its end.

About the Author

Joseph L. Bensinger is the author of eight previous poetry collections: *Man, Ships, and the Sea*; *Of Curses and Blessings*; *Journeys through the Tapestry*; *Beginnings and Ends*; *An Ekphrasis of Genesis*; *Layers of Consciousness*; *Echoes of Creation*; and *The Nested Soul*. His poems have appeared in *Haiku Journal*, *WestWard Quarterly*, *Ocean Magazine*, and *Mobius*. Before retiring, he worked as an electronics project engineer, a computer systems manager, and a teaching anthropologist. He now spends his time gardening and writing poetry.

Made in the USA
Monee, IL
05 November 2026

67381755R00049